DEADLIEST ANIMALS

GREAT WHITE SHARK

A Crabtree Branches Book

Amy Cullifor

Crabtree Publishing
crabtreebooks.com

School-to-Home Support for Caregivers and Teachers

This high-interest book is designed to motivate striving students with engaging topics while building fluency, vocabulary, and an interest in reading. Here are a few questions and activities to help the reader build upon his or her comprehension skills.

Before Reading:

- *What do I think this book is about?*
- *What do I know about this topic?*
- *What do I want to learn about this topic?*
- *Why am I reading this book?*

During Reading:

- *I wonder why...*
- *I'm curious to know...*
- *How is this like something I already know?*
- *What have I learned so far?*

After Reading:

- *What was the author trying to teach me?*
- *What are some details?*
- *How did the photographs and captions help me understand more?*
- *Read the book again and look for the vocabulary words.*
- *What questions do I still have?*

Extension Activities:

- *What was your favorite part of the book? Write a paragraph on it.*
- *Draw a picture of your favorite thing you learned from the book.*

TABLE OF CONTENTS

THE GREAT WHITE

It was the 1975 summer blockbuster film, *Jaws*, that not only changed Hollywood, but changed the world's awareness of sharks. The lead character was a giant great white shark.

It terrified a coastal community as it **preyed** on local boaters and swimmers. After seeing the film, many moviegoers found it hard to go back into the ocean.

GALEOPHOBIA MEANS FEAR OF SHARKS.

OCEAN PREDATOR

Perhaps no other animal is feared more than the shark. There are more than 400 species of sharks, and most are harmless. But some can be dangerous. The great white is the most well known and one of the most dangerous sharks.

Thanks to the movie *Jaws*, these ocean **predators** haunt the nightmares of many who venture into the water.

GREAT WHITE SHARKS ARE CARNIVORES, OR MEAT EATERS.

KILLER LOOKS

Great white sharks get their name from their coloring and their size. They are typically dark gray on top and white underneath.

They are part of a family of large sharks that have pointed snouts and large gill openings. Great whites are the largest predatory fish in the ocean.

GREAT WHITE SHARKS ARE KNOWN FOR THEIR ENORMOUS SIZE. THEY ARE TYPICALLY TWICE AS LONG AS HUMANS ARE TALL—ON AVERAGE 13 FEET (4 METERS). THE LARGEST GROW 20 FEET (6 METERS) LONG!

Great whites have muscular, torpedo-shaped bodies that are built for speed. Their powerful tails help them cruise through water with little resistance.

The most recognizable feature, however, is the shark's mouth. It's packed with razor-sharp, triangle-shaped teeth.

GREAT WHITE SHARK TEETH ARE SERRATED. THEIR SCIENTIFIC NAME, CARCHARODON CARCHARIAS, MEANS RAGGED TOOTH.

GREAT WHITE HABITAT

Great whites can be found just about anywhere in the world's oceans. They prefer cool coastal and offshore waters, especially where seals and other large prey live. Great whites often hunt near Australia, South Africa, and the U.S. coast of California.

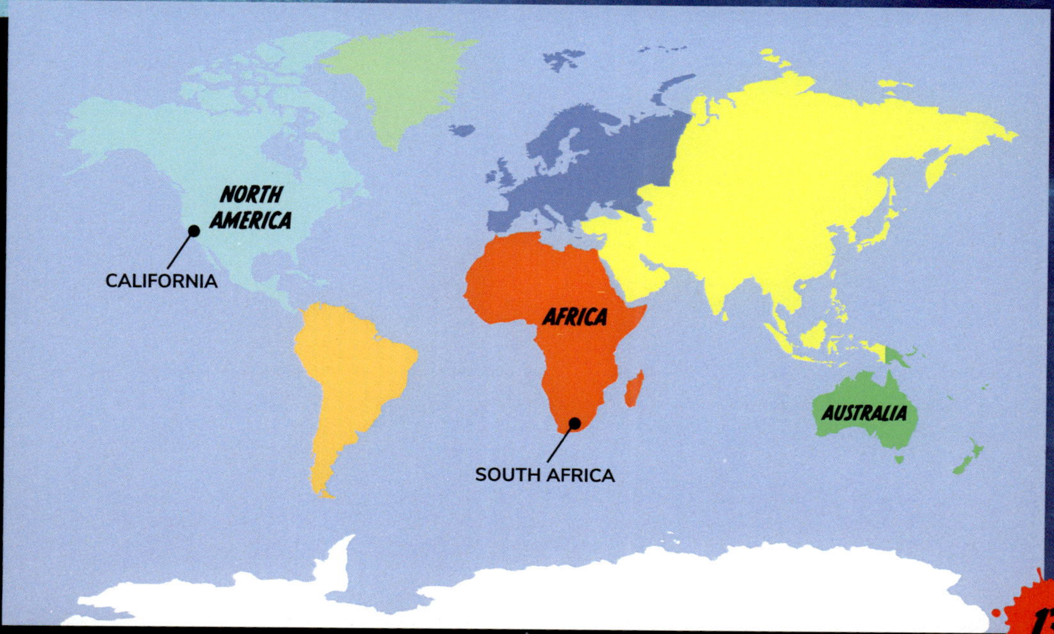
NORTH AMERICA
CALIFORNIA
AFRICA
SOUTH AFRICA
AUSTRALIA

OPPORTUNISTIC PREDATORS

Great white sharks are **opportunistic** predators. Their diet depends on what's available in the area. Commonly, they feed on large prey such as seals, sea lions, sea turtles, small whales, and fish. They even prey on other sharks.

THE LIFESPAN OF A GREAT WHITE IS HARD TO DETERMINE, BUT SCIENTISTS BELIEVE THEY CAN LIVE 50 YEARS OR LONGER.

Great whites are a **keystone species**. That means they help keep the oceans clean and healthy. One way they help is by **scavenging** the bodies of dead animals. Preying on marine animals also helps keep those populations in check.

From nose to tail, baby great white sharks are taller than the average ten-year-old kid!

Did You Know? Female white sharks don't lay eggs like some fish. They give birth to live baby sharks, called pups, that are nearly 5 feet (1.5 meters) long!

DEADLY WEAPONS

WEAPON 1—TEETH

A great white shark has five to seven rows of teeth. The first row has about 50 teeth, with multiple rows developing behind the first row. Great white teeth are perfect for gripping and ripping apart prey. When a tooth in the front row breaks off, a tooth in the second row moves forward to take its place.

A GREAT WHITE SHARK CAN GO THROUGH 20,000 OR MORE TEETH IN ITS LIFETIME.

WEAPON 2—SIZE

The biggest great white sharks are more than 20 feet (6 meters) long, and can weigh up to 5,000 pounds (2,268 kg). That's about the size of a pickup truck!

Weapon 3—Electroreceptors

Great white sharks are keen hunters, thanks to special sensing organs called electroreceptors. These tiny organs are located around the snout and lower jaw. They are jelly-filled pores lined with hair-like cells. The organs help detect electric currents in the water released by prey.

WEAPON 4–SPEED

Don't even think about trying to outswim a great white shark. They are among the fastest predators in the sea. In short bursts, great white sharks can reach speeds up to 35 miles per hour (56 kph). The fastest humans top out at 5 or 6 miles per hour (8 or 10 kph) in the water.

SHARK ATTACK!

In 1963, Rodney Fox was participating in a spearfishing competition. Without warning, a huge great white shark slammed into him. It grabbed him by his torso and refused to let go. The shark finally released Rodney after he jabbed his fingers into its eyes. But, the damage was done. Rodney suffered terrible injuries. The bite left him with huge open

wounds, shattered bones, and damage to his internal organs. Rodney was rushed to the hospital clinging to life. It took 462 stitches to close his wounds. Miraculously, Rodney recovered, although he will always bear the huge scar in the shape of the shark's mouth. Rodney says he still has a shark tooth embedded in his wrist. Despite the terrifying attack, he was back in the ocean spear-fishing a year later.

UP-CLOSE SHARK OBSERVATION

For scientists to study sharks, they need to get into the water. Often, they use shark cages to observe dangerous shark species like great whites. The heavy-duty metal cages offer a safe environment for up-close study.

Cage diving is also a popular activity in some places. Divers remain in the cage while the waters are **baited**, attracting sharks for observation. Underwater film crews use shark cages, too. That's how some of the incredible shark footage is filmed for the Discovery Channel's Shark Week programs that air every summer on TV.

RODNEY FOX HELPED PIONEER THE USE OF SAFETY CAGES TO PROTECT DIVERS WHILE IN THE PRESENCE OF DANGEROUS SHARKS. TODAY, HE IS A SHARK **CONSERVATIONIST**, AND ONE OF THE WORLD'S FOREMOST EXPERTS ON GREAT WHITE SHARKS.

People enter the sea millions of times every year. Yet, each year there are typically less than 100 shark attacks worldwide. Compare that to the more than 100 million sharks and rays that are harvested every year by the world's **fisheries**. It is easy to see why sharks have more to fear from people than the other way around.

SHARK ATTACKS RESULT IN ABOUT FOUR HUMAN FATALITIES ON AVERAGE EACH YEAR.

SUCCESSFUL PREDATOR

Great white sharks are one of the most **formidable** predators on Earth. Other than humans, their only natural enemy is the powerful killer whale. As a species, great whites are so successful, they have remained virtually unchanged for millions of years. That's why great white sharks are one of the world's deadliest animals.

SHARK ATTACKS MAKE THE HEADLINES, BUT THEY ARE QUITE RARE. SCIENTISTS BELIEVE MOST ATTACKS ON PEOPLE ARE ACCIDENTAL. THE SHARKS SIMPLY MISTAKE HUMANS FOR THEIR NORMAL PREY.

GLOSSARY

baited (BAYT-ed): Supplied with food

carnivores (KAR-nuh-vorz): Animals that eat meat

conservationist (kon-sur-VAY-shun-ist): A person who acts and supports the protection of wildlife and the environment

fisheries (FISH-ur-eez): Businesses that catch, process, and sell fish

formidable (FOR-mi-duh-buhl): Frightening

galeophobia (GAL-e-oh-FOH-bee-uh): Extreme fear of sharks

keystone species (KEE-stone SPEE-ceez): A species in an ecosystem that helps hold the ecosystem together

opportunistic (op-ur-TOO-nis-tik): To take advantage of an immediate situation

predators (PRED-uh-turz): Animals that hunt and eat other animals

preyed (PRAYD): An animal that hunted another animal for food

scavenging (SKAV-uhnj-ing): Searching through waste for food

serrated (SER-ay-tid): Having an edge like a saw

Index

Websites to Visit

https://kids.nationalgeographic.com/animals/fish/great-white-shark/

https://ocean.si.edu/ocean-life/sharks-rays/great-white-shark

www.natgeokids.com/uk/discover/animals/sea-life/great-white-sharks/

About the Author

Amy Culliford

Amy Culliford has a Bachelor of Fine Arts. She has worked as a classroom drama teacher and led after-school drama programs. She avoids sharks of any kind.

The author would like to thank David and Patricia Armentrout for their research and help on this project.

Crabtree Publishing

crabtreebooks.com 800-387-7650

Produced by: Blue Door Education for Crabtree Publishing
Written by: Amy Culliford
Designed by: Jennifer Dydyk
Edited by: Tracy Nelson Maurer
Proofreader: Crystal Sikkens
Production manager: Candice Campbell
Prepress technician: Katherine Kantor

Hardcover	978-1-4271-5413-2
Paperback	978-1-4271-5419-4
Ebook (pdf)	978-1-4271-5425-5
Epub	978-1-4271-5431-6
Read-along	978-1-4271-5437-8
Audio book	978-1-4271-5443-9

Printed in the U.S.A./CP052026

Library and Archives Canada Cataloguing in Publication

Title: Great white shark / Amy Culliford.
Names: Culliford, Amy, 1992- author.
Description: Series statement: Deadliest animals | "A Crabtree branches book". | Includes index.
Identifiers: Canadiana (print) 20210218924 | Canadiana (ebook) 20210218932 | ISBN 9781427154132 (hardcover) | ISBN 9781427154194 (softcover) | ISBN 9781427154255 (HTML) | ISBN 9781427154316 (EPUB) | ISBN 9781427154378 (read-along ebook)
Subjects: LCSH: White shark—Juvenile literature.
Classification: LCC QL638.95.L3 C85 2022 | DDC j597.3/3—dc23

Published in Canada
Crabtree Publishing
616 Welland Avenue
St. Catharines, Ontario
L2M 5V6

Published in the United States
Crabtree Publishing
347 Fifth Avenue
Suite 1402-145
New York, NY 10016

Photographs: Cover photo © Alessandro De Maddalena/Shutterstock.com, graphic splat on cover and throughout © Andrii Symonenko /Shutterstock.com,pages 4-5, 14, 15, 18, 19, 28, 29 ocean background © Alexmumu/istockphoto.com, full shark © Grisha Shoolepoff/istockphoto.com, shark fin © DigtialStorm/istockphoto.com, pages 6, 7, 26, 27 ocean background © vitacopS/istockphoto.com, page 6 shark © Howard Chen/istockphoto.com, page 7 shark © slowmotiongli/istockphoto.com, pages 8, 9, 11, 12, 13, 16, 17, 20, 21, 24, 25 ocean background © shevtsovy/istockphoto.com, page 8 shark © Whitepointer/istockphoto.com, fin © DigtialStorm/istockphoto.com, page 9 (top) © RamonCarretero /istockphoto.com, inset photo © solarseven/istockphoto.com, page 10 © atese/istockphoto.com, page 11 (top) ©Peter_Nile/istockphoto.com, (bottom) © BELOW_SURFACE/istockphoto.com, page 12 © William Buchheit /istockphoto.com, page 13 © USO/istockphoto.com, map © lukbar/istockphoto.com, page 14 shark © Peter_Nile/istockphoto.com, page 15 (top) atese/istockphoto.com, (center and bottom) © USO/istockphoto.com, page 16 © USO/istockphoto.com, page 17 shark © Geerati/istockphoto.com, girl © loco75/istockphoto.com, page 18 © Mark Kostich/istockphoto.com, page 19 shark © ELizabethHoffmann/istockphoto.com, jaws © geckophoto/istockphoto.com, truck © Vladimiroquai/istockphoto.com, page 20 both sharks © Alessandro De Maddalena /istockphoto.com, page 21both sharks © RamonCarretero /istockphoto.com, pages 22-23 © Willyam Bradberry/Shutterstock.com, page 24 © atese /istockphoto.com, page 25 © izanbar/istockphoto.com, page 26 © rypson/istockphoto.com, page 27 (top) atese/istockphoto.com, (bottom) © Koichi Yoshii/istockphoto.com, page 28 © Tsvyatko Terziev/istockphoto.com, page 29 (top) © MogensTrolle/istockphoto.com, (bottom) © Nuture/istockphoto.com

Library of Congress Cataloging-in-Publication Data

Names: Culliford, Amy, 1992- author.
Title: Great white shark / Amy Culliford.
Description: New York : Crabtree Publishing, 2022. | Series: Deadliest animals - a Crabtree branches book | Includes index.
Identifiers: LCCN 2021022064 (print) | LCCN 2021022065 (ebook) | ISBN 9781427154132 (hardcover) | ISBN 9781427154194 (paperback) | ISBN 9781427154255 (ebook) | ISBN 9781427154316 (epub) | ISBN 9781427154378
Subjects: LCSH: White shark--Juvenile literature. | Dangerous animals--Juvenile literature.
Classification: LCC QL638.95.L3 C85 2022 (print) | LCC QL638.95.L3 (ebook) | DDC 597.3/3--dc23
LC record available at https://lccn.loc.gov/2021022064
LC ebook record available at https://lccn.loc.gov/2021022065